John Townsend is a Franco-British writer, musician, composer, and artist who has spent most of his life in France. After graduating from Exeter University with a degree in sociology, he worked for a couple of years as a Human Resource specialist in London before co-founding a successful rock band which had a number one hit in France. After the band broke up, John decided to stay in France and went back into HR. After working in managerial positions in France, Switzerland and the USA, he founded the Master Trainer Institute on the French-Swiss border. On retirement he began a successful career as a painter of seascapes. He and his wife now live in Brittany where he recently digitised this book of his 1972 journey of discovery.

JOHN TOWNSEND

BRIDGE HUNT

A Journey Through Switzerland to Rediscover
the 42 Forgotten Bridges of Robert Maillart

AUSTIN MACAULEY PUBLISHERS™

LONDON • CAMBRIDGE • NEW YORK • SHARJAH

A CIP catalogue record for this title is available from the British Library.

ISBN 9781398451162 (Paperback)
ISBN 9781398451179 (Hardback)
ISBN 9781398451186 (ePub e-book)

www.austinmacauley.com

First Published 2024
Austin Macauley Publishers Ltd®
1 Canada Square
Canary Wharf
London
E14 5AA

To Fiona for her patience and guidance, to Robert Maillart and his family for inspiration, and to Tom Holt for lending me his camera in 1973.

Table of Content

Foreword

This book was written in 1972 when I was 28 years old and living in Ferney-Voltaire on the French/Swiss border. I typed it up finger by finger on a clickety-clackety old-fashioned typewriter on sheets of coloured card, glued on the photos and then punched the pages and bound them with an Ibico mechanical spiral binding machine.

Although I only really wrote it for myself as a sort of autumn/winter project, I did take the manuscript to a local 'art' publisher in Geneva who promised to schedule it for publication for a small and specialised market. Then I was offered a promotion to the US headquarters of the company I was working for and the book was packed up and forgotten in the move to Connecticut. Since my return to France in 1976, it has remained in various cupboards having survived 6 further house removals – with occasional dustings, repeated vague but unfulfilled promises from that original Swiss publisher and the compiling of one hastily made photo-copy for an interested but budget-less professor of Civil Engineering at the ETH in Zurich.

For these past 50 years Robert Maillart's reputation as a pioneering genius in the art of bridge-building has spread and more and more articles and books on his work are appearing every year. His Salginatobel Bridge has been named a Swiss heritage site of national significance and in 1991 it was designated an International Historic Civil Engineering Landmark by the American Society of Civil Engineers.

However, as far as I know, there is still no book which 'visits' all of the bridges Maillart ever built in Switzerland (plus one just over the border in France). In 1972 only the Tavanasa Bridge had disappeared, destroyed by an avalanche in 1927.

Now, in 2024, Google shows me that the French Bridge over the Arve in Vougy has been transformed beyond recognition, the Klosters Bridge has been replaced by an ugly 'modern' stucture and the Val Tshiel Bridge in Donath has been bypassed and now stands forlorn beside its successor.

Amazingly, in November 2022 while fact checking my original 1973 manuscript, I was reading through David P. Billington's 1979 book "Robert Maillart's Bridges – the art of engineering" and discovered that I had missed his very first bridge! It had never been mentioned either by Maillart's family or on any of the documents available to me at the time.

It seems that, as a young 22-year-old graduate trainee, Maillart had helped to design and supervise the construction of a small railway bridge near Morges in the canton of Vaud. Rather than tamper with my original manuscript I have included this 43rd. bridge in the 'Afterword' – along with Billington's text and a recent photo.

Now, at 79 years old, I have no longer the energy to undertake a new journey to revisit those bridges still standing/functioning but, with today's technology of OCR and photo scanning at my fingertips, I have decided to rediscover my 1972 journey of rediscovery and relook the manuscript! Thanks to Google Maps I have also been able to provide accurate coordinates for each of the bridges. Happy bridge hunting!

John Townsend, Brittany, France 2024

The Reason Why

It was late September in the year of 1972 when I first had the idea of writing this book. Although I did not realise it at the time, as I drove up from a holiday in Italy to Switzerland and as the first patches of autumn snow started to appear beside the winding alpine road, the Indian summer I had left down in the valley of Lake Como marked the centenary of the birth of Robert Maillart.

Since then, many of my friends, on hearing of my project have asked me, "But who was Robert Maillart?"

I'm not really surprised because my own superficial acquaintance with the great but neglected Swiss engineer had only begun in that same year in which, had he lived, he would have been 100 years old. Living near Geneva, I had decided to take my belated 1972 holiday on the Italian Isle of Elba. On the long drive home, sunburnt and rested, I decided to make a detour and, climbing back into Switzerland from the blue crystal lake of Como I headed for St. Moritz.

The reason for this detour was in the glove compartment of my car. A book which I had bought just before leaving on holiday written by Derrick Beckett and simply entitled "Bridges".

Browsing through the travel books, I had been struck by a photograph on the back cover of this rather special work. In the midst of mountain ravines and evergreen forests hung a silver thread of concrete. I am not an architect or an engineer and know almost nothing about stresses and moments, but bridges and all they represent had always fascinated me. My interest aroused, I flicked through the pages trying to find where such a beautiful bridge could be situated and who had built it.

Derrick Beckett had written a chapter entitled "The Salginatobel Bridge" and there to my amazement I discovered that the bridge was hidden away in one of Switzerland's beautiful but remote and misty valleys. There too I discovered the name of the genius who had conceived and constructed this wonderful work of art – a certain Robert Maillart (1872–1940). Along with the picture of the Salginatobel Bridge, which had fired my imagination, was a picture of one of Maillart's first bridges (I later discovered it was his second) built over the river Inn at Zuoz, in the Swiss Canton of Graubünden in the year 1902.

As I had gazed at the faded blotchy photograph, showing the scaffolding still in place, I could only wonder how such a slender modern structure could have been built as early as 1901 and had resolved to see if it was still standing. I couldn't resist the hope that I had stumbled upon a Swiss Telford or Brunel.

Leaving St. Moritz on that September afternoon I followed the winding River Inn along the valley. Soon I was entering the picturesque little village of Zuoz and my heart quickened. Could this little bridge, possibly built to bring the cattle and the timber across from the other side of the valley, still be standing? Or had the townsfolk, in their relentless progress towards a modern Switzerland, destroyed the bridge and built a new and wider way across the river.

They had obviously felt that Maillart's visionary work was quite worthy of its place on their river because, as I turned on impulse off the main road and wound down between the little villas and farmyards, the buildings suddenly gave way to green meadows stretching down to the rushing waters of the River Inn. There, arching like a taut bow across the river was Maillart's bridge.

I spent some time contemplating the bridge, as one might a painting, took my first, unprofessional photograph and having quenched my thirst and wiped my boots (the towpath was quite muddy from the early autumn rain) I set off once more towards one of the distant alpine passes.

I remember that I had the car radio on and it was interfering with my thoughts so I switched it off as I started the tortuous climb up to the Flüela Pass. I was on my way to find the Salginatobel Bridge at Schiers.

Maillart's Second Bridge (Zuoz, 1902) and my first photo!

The snow about me was already deep and the mist seeping lower onto the pass so that I had to put on my headlights. I don't know why the idea of writing something about Maillart came to me then since, for all I knew, there might already be any number of books written about him. In fact, I told myself, if he really were Switzerland's own Brunel or Telford, I might as well forget it.

Even so, the thought nagged me as I drove down into the sunlight at Davos. Maybe it was worth a little more research.

Several hours later however, I had given up all hope of finding the Salginatobel Bridge. Tired and disappointed, I searched the pages of Derrick Beckett's book for the last time, vainly trying to find a reference as to where the beautiful silver thread could have gone. All I found was, 'The Salginatobel Bridge near Schiers in Switzerland' and, armed only with a touring map, I just couldn't locate it. It was getting late and there was no hope of buying a more detailed map.

Perhaps the bridge had been replaced when the new road from Landquart to Davos had been built. I had driven at least 10 kilometres up the only valley which cut down into Schiers, but to no avail.

The roadway wound up into the mist and was so narrow there was not even room for two cars. I couldn't believe that such a beautiful bridge could have been built to serve what was virtually a farm track. So I gave up and started the long, cold, night-drive home to Geneva.

If only I had known that I had come, map-less and dispirited, to within a kilometre of the hidden and forgotten masterpiece which, despite its unique aspect of grandeur, is only just wide enough to take a tractor!

Sir Edmund Hilary fought against the hardest odds to climb Mount Everest "because it was there" and paradoxically I continued my search for the mysterious Salginatobel Bridge simply because it wasn't there – or didn't seem to be.

Of course, my overdramatising of the situation was brought abruptly to an end when, a few days after my return to Geneva, I managed to get hold of a 1/25000 scale map of the Schiers area and found my lost bridge only a kilometre or so from where I had stopped and turned round.

At that moment, I vowed to return and photograph the bridge at the same time promising myself to find out more about the enigmatic Robert Maillart.

My project, from its birth on the Flüela Pass to the completed manuscript took just over a year of part time work. Exciting, enjoyable and healthy work with over 400 photographs taken and many hours of browsing through literature, talking to Maillart's relatives, poring over maps, climbing, walking and scrambling and now, at last, writing. After my "discovery" of the Salginatobel Bridge, I paid a visit to the biggest bookstore in Geneva and enquired after books written by or about Robert Maillart. Like any researcher, however meagre, I secretly hoped there would be none.

However, after a long moment consulting lists and digging into reference books the triumphant assistant announced the existence of "Robert Maillart" by Max Bill, the well-known Swiss artist and sculptor – the book which became my bible in my search for forgotten bridges.

My initial disappointment gave way to pleasure during the next days, as I buried myself in the world of Maillart's bridges (and as I discovered, other reinforced concrete constructions) 34 of them were described and documented by Max Bill with original photographs taken at the time of construction (35 if we include the Tavanasa Bridge destroyed by an avalanche in (1927).

Unfortunately, it had been one of the most beautiful bridges Maillart ever built.

I must admit that I got very lost with many of the technical descriptions and data (I still am) but my admiration for Maillart, the artist, grew with every photo of the elegant space-defying structures. I learned that Maillart was one of the pioneers of reinforced concrete. As a complete layman, I read with interest of his experiments and theories on how to thread metal bars through concrete to strengthen it – especially since the resulting spans seemed so thin and tenuous.

The Saginatobel Bridge near Schiers

With examples of massive ugly concrete slabs all around me as I travel modern Europe in my job, I marvelled how Maillart had managed to conquer his spaces so delicately – often giving the impression that his bridges were suspended by an invisible wire. Or maybe that they had grown out of the very soil (like the Winterthur Bridge) and crossed the river themselves like a supple willow branch.

This slenderness and grace, however, did not apparently impress the Swiss authorities in his day since, as my later journey showed me, his bridge projects were for the most part relegated to hidden valleys or farm track railway crossings. So hidden and forgotten are many of his structures that they are almost impossible to find, even with the aid of large-scale maps. But we cannot blame his patrons for their scepticism. Maillart was a true pioneer and had to suffer the slings and arrows of frustration known to each and every innovator in any field of human achievement. I shan't dwell on his lack of recognition during his lifetime; only try and ensure like Max Bill (but as a layman) that he is not forgotten.

A few days later, I set off for the public library in Geneva to find out whether anything else had been written on Maillart. I discovered one or two articles listed from ancient construction magazines, all written in German, and asked the librarian if there were no books on Maillart. She disappeared into the depths of the archives and returned thirty minutes later clutching a thin, square-shaped book. It was an earlier edition of Max Bill's book which I had at home.

Since nothing else seemed to have been written which could tell me more about the mysterious Robert Maillart, I decided upon a very simple course of action. In my office, the next morning, I opened the Geneva telephone book at M and found three Maillarts. The first of the three numbers did not reply but on trying the second, a young man answered the phone. I did not quite know how to begin tactfully so I just asked him outright if he was a relation of the "famous" Swiss engineer Robert Maillart who had died in 1940. He hesitated and then told me he was a distant relative. However, when pressed he admitted that he knew next to nothing about Robert Maillart but did give me the address and telephone number of his uncle in Bern who could tell me more. I hung up and thanked my good fortune. I was on the trail at any rate.

Mr Alfred Maillart in Bern was extremely helpful and told me that if I really wanted to know something about the late Robert Maillart then my best contact would be Mrs Blumer-Maillart in Zurich – Robert's daughter. Apparently, she was at that very moment preparing a trip to the USA where she would be representing her father at a conference on his work in Princeton University, New Jersey.

When Mrs Blumer-Maillart returned from Princeton, we were able to get together to discuss my crystallising project.

I am so grateful for the assistance which has been given to me since these first few timorous phone calls. I have discovered that the family is trying to help those who wish to ensure that Robert Maillart's name is remembered and that his work be appreciated in all its fullness.

I have been given copies of articles written in various journals and reviews and discussed the projects mounted by Princeton and Cambridge universities to bring Maillart's work closer to the attention of architects and engineering students. I have spoken to David Billington, Professor of Civil Engineering at Princeton and compared notes. David's work on Maillart, in the true tradition of the engineering scholar, has left me in awe and, I admit, still rather lost.

At every turn, I have felt immense satisfaction that so much work is being done by these dedicated people to preserve the genius of Robert Maillart and would have given up my task from pure technical ignorance, if not for two things.

Firstly, I feel that bridges are more than mere technical achievements balancing the physical forces of nature against man's ingenuity. To me a bridge, as a concept, is a symbol of civilisation and as a finished work becomes a form of artistic expression. A symbol because it represents sociologically one of the noblest attributes of mankind – "rapprochement" or bringing closer together. No bridge (except in times of war or civil strife) was ever built which only allowed people to go one way. In simple terms a bridge represents the urge to conquer natural barriers in order to bring people closer together.

At least this is the feeling I get when I contemplate a bridge and, as a non-architect and hopeless in technical matters, this book is about my feelings when I look at Robert Maillart's work.

As a form of artistic expression, Maillart's bridges represent one man's idea of how to span spaces in the simplest and most beautiful way. Brunel's Clifton Suspension Bridge and the Golden Gate Bridge over San Francisco Bay represent other men's artistic solutions to the problem. What has struck me in my pilgrimage to the remote valleys and lonely lakes of Switzerland is how Maillart remained true to his ideals and with each bridge he built created the most aesthetically satisfying structure possible given the technical difficulties or administrative limitations – even in remote valleys where the most tasteless and ugliest of structures would have satisfied his patrons.

The second reason why I have continued my project in the face of far worthier and erudite work at present being carried out is that, to the man in the street, Maillart is still unknown. Who has not heard of and admired the work of Eiffel, Le Corbusier, Picasso and Henry Moore? Why then should the layman not become acquainted with Robert Maillart?

At the risk of being considered a shallow amateur, I also nurture a hope that my project may even be useful to the engineers, architects and students who are already aware of Maillart's work.

As far as I know, this book shows every bridge (outside Russia) ever conceived and constructed by Robert Maillart (except the Tavanasa Bridge destroyed by an avalanche in 1927 and one of the two Wägitaler See bridges which was replaced when the road round the lake was widened). The photographs arc in colour showing the bridges as they are now in 1973. To my knowledge no other such collection exists.

Madame Blumer-Maillart has appreciated my interpretation of her father's work and since the beginning of our acquaintance has recognised the charm of such a non-technical book. The following chapter on Robert Maillart's life is based entirely on her memories of her father and his fascinating vision. It has been compiled from cassette tapes on which she has spent so much time and effort to record and from conversations we've had, where we tried to explore Maillart the man rather than Maillart the engineer.

This is how it all began. All that remained was to go and find the bridges.

By an immense stroke of fortune, I discovered that the Swiss Association of Cement Manufacturers had, on the occasion of Maillart's centenary, decided to publish a small map of Switzerland showing the location of each bridge. I later discovered that the map is somewhat imprecise as to the exact placement of some of them but it was a wonderful help. At least all I had to do now was to plan my route, load up the car, borrow a good camera (my faithful Instamatic could unfortunately not give me the quality I wanted) and disappear into the beautiful loneliness only Switzerland can offer.

John Townsend, Ferney-Voltaire, France 1973

"Ich baue für die Ewigkeit und wenn es vorher abekeit, so dut's mir fürchtig leid"

(I'm building for eternity but if one collapses before that, then I'm frightfully sorry)

Robert Maillart (1872-1940)

Robert Maillart (1872–1940)

Robert Maillart was born on February 6th 1872 in Berne, the capital of Switzerland. His father, a Belgian banker, had just established himself in this elegant city which was the home town of his young Swiss wife. Fraulein Küpfer, as she had been known before her marriage, had strong ties with Berne. In fact, her grandfather had been "Rittmeister" of the city charged with the management of the official stables which provided the horses for ceremonial occasions.

Robert was the third of the family's four boys and as he grew up he probably did not realise that he had such an illustrious ancestry. His modest father did not choose to call himself "Le Baron de Maillart" even though he was officially allowed to use the title. In fact, the family can trace its genealogy right back to the 10th century when they were important figures in the European feudal aristocracy.

The history of the Maillarts is long and full of daring deeds and quaint anecdotes. One fascinating chapter in their story concerns one Colley Maillart, known as Colin to the locals. Colin Maillart was tragically blinded in one of the many feudal battles which raged during the middle ages but, back in the peaceful grounds of his castle, he liked to amuse the village children by playing a very real version of "Blind Man's Bluff". The name of this children's game in French has remained "Colin Maillart" to this day.

Robert himself could have adopted the title of Baron but inherited his father's modesty and remained plain Herr Maillart throughout his life.

In 1889 Robert was 17 years old and had already graduated from high school in Berne. He was too young to enter the engineering school in Zurich's Eidgenossisches Polytechnikum and decided to go to the Horlogerie school in Geneva.

After a pleasant year studying the art of the watchmaker in the peaceful French-speaking city Robert signed on for the four-year course in civil engineering at Zurich. He excelled in most of the subjects on the tough curriculum but, ironically, one of his lowest grades was in bridge design! (See the copy of his Graduation Certificate below).

EIDGENÖSSISCHES POLYTECHNIKUM

ZÜRICH

ABGANGS-ZEUGNIS

Herr Robert Maillart

von Bremgarten, Kt. Bern, geb. 1872

besuchte vom Oktober 1890 bis März 1894

als regelmässiger Schüler der ersten, zweiten, dritten, vierten Jahreskurse der

INGENIEUR-SCHULE.

Während seines Aufenthaltes an der Anstalt hat er die nachbenannten Fächer bewohlt und es sind ihm für seine Leistungen in den obligatorischen erteilt worden (2 bedeutet)

A. Unterrichtsfächer mit obligatorischem Charakter	Lehrer	Noten
Analytische Geometrie	Professor Dr. Geiser	5½
Mechanik mit Uebungen	Professor Dr. Herzog	5½
Physik	Professor Dr. Weber	6
Technische Geologie	Professor Dr. Heim	6
Baukonstruktionslehre	Professor Recordon	4½
Baukonstruktionszeichnen	Derselbe	5½
Planzeichnen	Professor hon. Bosch	6
Kartenzeichnen	Derselbe	6
Topographie	Professor Dr. Pachler	5½
Feldmessen	Derselbe	5½
Geodäsie	Derselbe	5½
Technologie der Baumaterialien	Professor Tetmajer	6
Baumechanizlehre mit Konstruktionsübungen	Professor Prüs	4½
Praktische Hydraulik	Professor Plingner	5½
Theoretische Maschinenlehre	Derselbe	2
Graphische Statik mit Konstruktionsübungen	Professor Ritter	5
Steinschnitt mit Konstruktionsübungen	Derselbe	5
Steinerne, hölzerne und eiserne Brücken	Derselbe	4½
Konstruktionen dazu	Derselbe	4½
Die Fundationen	Professor Zschokke	4½
Konstruktionsübungen dazu	Derselbe	4½
Schiffbarmachung der Flüsse und Kanalbau	Derselbe	5½
Konstruktionsübungen dazu	Derselbe	4½
Entwässerung und Bewässerung	Derselbe	6
Bau am ?		6

After graduation, Maillart worked for three years with Pumpin and Herzog, a private civil engineering company in Berne. In 1897 he moved back to Zurich and worked for the municipality. It was during this time that he designed his first bridge, the Stauffacher Bridge in Zurich. This first "Maillart system" bridge was hidden under a mass of ornamentation dictated by the mode of the time. Even so, it has a lighter look about it than many of the classical bridges being built at the turn of the century.

In 1901 Maillart designed and built his second bridge at Zuoz near St. Moritz (the sight of which had launched the idea for this book!) During its construction the young engineer would often visit the site to supervise the work for his concepts were new and the construction had to be precise. He always stayed at the Enghadina Hotel in Zuoz and one night, while seated at the high table, a dark-eyed

Italian girl on holiday with her parents caught his attention. It was love at first sight and the pair were married within a few months.

After what was then eight years as an employee, Maillart decided in 1902 to work for himself and founded his own company in Zurich for design and construction. Maillart et Compagnie prospered and rapidly expanded with projects being carried out as far away as Spain and Russia.

Between 1902 and 1914 Maillart built eight bridges, all of them in Switzerland. However, his other engineering business in Russia was flourishing and in 1914 he decided to take his wife and three young children there on an extended visit. His daughter Marie-Claire, who has helped me so much with this book, remembers very clearly the family's stay in Russia.

The summer of 1914 passed idyllically for the young Maillart children, installed in a comfortable apartment in the Hotel Kevitch in Riga. They went for long walks on the vast, sand dunes of the Baltic and shared picnics in the delightful woodland behind the town. But winter came, the sea froze on the sand and rumblings of war further south told them that it would be difficult to return home. During this time, Maillart was working on twenty or so small railway bridges on the line leading to St. Petersburg.

Later on in September 1915 one of his sons had to go back to Switzerland via Finland because of the war in Europe and the family moved to Harkov where Maillart was building a large warehouse and factory complex. Marie-Claire remembers the immense "chantier" where a thousand workers were employed for the concrete alone and a little train brought supplies up to the work site.

In the winter of 1916, Maillart worked on a project for a bridge in St. Petersburg. It seems that the bridge was never actually built while Maillart was in Russia but, curiously enough, a bridge exists in St. Petersburg today which bears the unmistakeable mark of the Maillart system (photo on next page).

Мало-Крестовский мост.

It is still not known when this bridge was constructed but it is very possible that the Russians used Maillart's original plans to build the bridge after his flight back to Europe.

In the same year of 1916 Maillart's wife fell ill and in the month of August she was hospitalised and died. Robert remained in Harkov with his children until the revolution broke out in 1917. In the first outbreak his life was spared, despite some terrible moments, because he was a Swiss national.

Marie-Claire remembers the frightening brutality of the Bolsheviks including the firing squads, the drunken murders and the drownings in the river which ran through Harkov. In the winter, the Germans came to the town and captured the garrison.

Times were easier under the German occupation until the armistice which was signed in 1918 and in December of that year a faithful employee came to the apartment one icy night and warned Maillart of the imminent, forced liquidation of his business. That fateful midnight, still numbed by the loss of their mother, the children dressed in the bitter cold and Robert and his family fled the town on the first available freight train.

After many an adventure, moving slowly south through Odessa to Greece and the Mediterranean, the Maillarts arrived back in Geneva having lost everything they owned.

Never again was Maillart to have his own construction business. From 1919 when he opened a small office in Geneva until his death in 1940, he concentrated on design. During the twenties he started to break away from the more conventional bridge projects and in 1930 reached his zenith with the designs for the Salginatobel and Landquart Bridges.

From then on, he completed over thirty bridges of incredible originality and artistic design. For the most part, these are the bridges which have established him as a pioneer and a genius.

During these years, one of Maillart's closest friends was Mirko Ros whose job, among other things, was to test the bridges which Robert had designed. The photo on page 27 shows a contented Maillart with his friend after the successful testing of the Val Tschiel Bridge in Donath.

Nevertheless, despite all the successful tests, none of the Swiss municipalities would accept his designs for prominent town locations. Even so, his reputation was growing among architects and engineers, and in 1936 he was elected an honorary member of the Royal Institute of British Architects.

This nomination may have surprised some of the cantonal authorities at the time but, looking back, no one can be surprised at the interest which is at present being shown all over the world wherever students and architects gather together to discuss art and engineering.

My task was to bring Maillart's genius to the attention of the man in the street and, with this in mind, I was ready to set off on my BRIDGE HUNT.

Robert Maillart with his wife and three children circa 1912

*Maillart with friend and colleague Mirko Ros after the
successful testing of the Val Tschiel Bridge in Donath*

The Bridges

Not all Maillart's bridges are as elegant and spectacular as the Salginatobel. Like all engineers, Maillart used various systems of construction and various materials to solve his problems. Simple and continuous beam bridges in reinforced concrete and masonry arches were added to the beautiful concrete hinged-arch and arch-rib bridges as part of a rich repertoire of space-defying solutions.

Least eye-catching are the masonry arch and beam bridges but I have included them one by one as I discovered them. The order and the numbering of the bridges follows the sequence of my journey. No attempt has been made to group the bridges by style or year of construction; no effort to trace the evolution of Maillart's creative genius. I am not qualified to evaluate this aspect of his work and simply appreciated every bridge for itself.

However, to assist the student and the engineer, I have drawn up a table on the following two pages which lists the bridges in the chronological order of their construction.

Maillart's beautiful footbridge in Winterthur

The Bridges of Robert Maillart
1896–1940

BY YEAR OF CONSTRUCTION

43. In "Afterword". PAMPIGNY – The Veyron Bridge 1896

32. ZURICH – The Stauffacher Bridge: 1899

24. ZUOZ – The Inn Bridge: 1901

29. OBERBOREN – The Thur Bridge: 1903/4

26. ST. GALLEN – The Steinach Bridge: 1903

27/28. AACH – The Railway Bridges: 1907

21. WATTWIL – The Thur Bridge: 1909

34. RHEINFELDEN – The Rhine Bridge: 1909

33. LAUFENBURG – The Rhine Bridge: 1911

36. AARBURG – The Aaar Bridge: 1911/12

16. IBACH – The Muota Bridge: 1912/13

2. VOUGY (France) – The Arve Bridge: 1920

19. INNERTHAL – The Schrabach Bridge: 1924

20. WÄGITALERSEE – The Ziggenbach Bridge: 1924

3. CHATELARD – The Aqueduct: 1925

25. DONATH – The Val Tschiel Bridge: 1925

39. BERNE – The Lorraine Bridge: 1928/30

22. SCHIERS – The Salginatobel Bridge: 1929/30

23. KLOSTERS – The Landquart Bridge: 1930

9. ADELBODEN – The Spitalbrucke: 1930/1

8. LADHOLZ – The Engstligen Bridge: 1930/1

6. SCHANGAU – The Hombach Bridge: 1931

7. SCHANGAU – The Lüterstalden Bridge: 1931

15. NESSENTAL – The Triftwasser Bridge: 1932

10. HABKERN – The Traubach Bridge: 1932

11. HABKERN – The Bohlbach Bridge: 1932

40. SCHWARZENBURG – The Rossgraben Bridge: 1932

41. SCHWARZENBURG – The Schwandbach Bridge: 1933

30. FALSEGG – The Thur Bridge: 1933

13. INNERTKIRCHEN – The Aare Bridge: 1934

31. WINTERTHUR – The Toss Footbridge: 1934

35. LIESBERG – The Birs Bridge: 1935

37. HUTTWILL – The Railway Bridge: 1935

42. TWANN – The Waterfall Bridge: 1936

1. GENEVA – The Arve Bridge: 1936

12. ZWEILÖTSCHINEN – The Gündlischwand Bridge: 1937

38. BERNE – The Railway Bridge: 1938

14. WYLER – The Gadmerwasser Bridge: 1938

17. ALTENDORF – The Railway Bridge: 1939

4. LAUBEGG – The Simme Bridge: 1939/40

5. GARSTADT – The Simme Bridge: 1939/ 40

18. LACHEN – The Railway Bridge: 1940

The Journey

It was five-o-clock of a May morning and the sun was just breaking into the sky over the Alps, as I packed the car with the equipment I would need for the journey. The previous week had been spent planning the best possible route I would need to take in order to visit all 42 of Maillart's bridges in the time available. I had managed to borrow a Pentax camera from friend Tom Holt and had 10 rolls of Ektachrome film ready for the Bridge Hunt. My instamatic would stay with me as a standby if I missed any shots, since I do not consider myself a real photographer.

I had an exciting "safari" feeling as I loaded the car with maps, a cassette recorder for my notes and scruffy clothes for the inevitable scrambling and soakings.

I had already made one or two excursions with my camera-owning friend. We had found the five bridges nearest Bern and captured the fullness of the Swiss autumn colours. I had learned from experience that bridge hunting can be a dangerous sport if you want to get just the right angle for the composition you're seeking.

In passing I should mention here that these five bridges have been added on to the end of my journey for the sake of continuity although they had been photographed the previous autumn.

As I drove quietly into downtown Geneva, the sun was already picking out slanting reflections on the calm waters of Lake Leman and a few early morning workers were strolling to bakeries and building sites, pausing to take in the peaceful early morning calm that has made Geneva one of the most beautiful cities in the world.

My journey had begun.

1. Geneva – The Arve Bridge (1936)

The road from Champel drops down into the end of the world. The Bout du Monde area is now built up into a sports and recreation area with arenas and soccer pitches and here, gracefully spanning the Arve river, stands one of Maillart's finest bridges. This bridge takes travellers out of the heavily populated suburbs of Geneva and, in a single leap, transports them across to the delightful farmland nestling under the Salève mountain and into the little village of Vessy.

I stopped at the side of the road and walked down to the riverside in the early morning sunshine. Truly, everything which Maillart wanted to express in his bridge-building is to be found in this beautiful three-hinged arch with its slender supports.

It is said that he exploited the structural limits imposed by the cantonal authorities to the maximum. I could almost hear the tuts and admonitions of the inspectors as they regarded the incredibly delicate struts which hold the solid bridge and heavy roadway in its flight across the river.

I must admit that I cannot bring myself to blame their reticence because it is only now that Maillart's absolute faith in his interpretation of the behaviour of reinforced concrete under stress can be justified.

Very few of his other bridges express so visually his disregard for massive masonry and his battle for maximum simplicity.

But can the ascetic beauty of this and other Maillart bridges be explained simply by his quest for economy of material? Is the essence of his artistic creation due only to the fact that he made every kilo of concrete, every centimetre of reinforcement do all the work of bearing the stresses and strains inherent in a bridge? I think not.

Calculations alone do not produce works of art. If we gave a full breakdown of all the various elements of stress involved in spanning the Arve at Vessy to an average civil engineer, I cannot believe that he would produce Maillart's beautiful bridge.

(Map Coordinates: 46.181063, 6.159788)

2. Vougy – The Arve Bridge (1920)

My next stop provided a complete contrast to the bridge over the same river which I had left 40 minutes before. Here at Vougy (just over the border in France) stands a lovely, old, and mellowed bridge built with 3 arches onto two solid piers. It is not a spectacular sight but I could not help thinking how well it fitted into the surroundings.

The arches are light and airy and not filled in as with his Rhine Bridges. It gives the impression of a graceful, well-preserved old grandmother – an impression which is enhanced by the delightful railing running along the sides and the little fluted kerb stones which separate the roadway from the narrow pedestrian passages.

How sad to see the motorway earthworks scarring the countryside all around. I managed to exclude them from the photo but the modern giant bulldozers and the workers' huts are getting dangerously near. Roadman spare that bridge! Only two of his creations have disappeared (one by natural disaster). Let us hope Vougy will not be the third.

In 2019 a photo sent to me by the Mairie of Vougy shows that the bridge is still there but unfortunately renovated, widened and rendered unrecognisable.

(Map Coordinates: 46.070029,6.497106)

3. Chatelard – The Aqueduct (1925)

After Chamonix the road gets narrower and the air gets colder as you climb to the cleft in the Alps known as the "Col des Montets". The overpowering beauty of Mont Blanc dominates the drive and it is only as you come down into the valley of the Barberine that the vista is left behind. Just after the customs post and around the bend back into Switzerland there it is, bold and magnificent as ever. Aqueducts to me had always meant ugly honeycombs of rusty iron or boring interminable arches. How novel a solution Maillart found to carry water across water. One can immediately see his style in the aqueduct often repeated in his "moulded" railway crossings.

As I stood on the roadway that runs along the side of the Eau Noire and gazed up at the ageing aqueduct, a man suddenly appeared and walked across the top. I would have been less surprised to have seen a tugboat and, clambering up the steep escarpment, found what looks to be a perfectly normal footbridge.

In fact, the water runs through the middle in a kind of concrete tunnel. As I drove away, I thought how unfair it was that even on a large-scale map, this unique structure appears only as a thin blue line across the road and the river.

(Map Coordinates: 46.0590264, 6.9603857)

4. Laubegg – The Simme Bridge (1939/40)

Of all Maillart's bridges, this one at Laubegg was the most disappointing and I shall pause here only to indicate its position.

I took the road from Aigle up through Chateau d'Oex, Saanen and Zweisimmen and only a few miles further on came into the hamlet of Laubegg where this non-descript bridge spans the Simme.

From an engineering and structural point of view the Laubegg Bridge is one of the simplest Maillart ever built, and I must say it holds no real attraction for me.

(Map Coordinates: 46.5877177, 6.3815462)

5. Garstadt – The Simme Bridge (1939/40)

Only a few hundred yards down the road, the Garstadt Bridge restored my confidence in Maillart! Here is his second to last bridge where he provided a beautifully-elegant bridging solution by crossing the river obliquely so that the approach roadways curve gently in a well-thought-out flowing pattern.

My picture does not do justice to the futuristic lines of the bridge as it leaps out from the banks to meet at an incredibly thin hinge in the middle. The daringness of Maillart's vision is accentuated by the fact that he used bold straight lines rather than a curving arch. This is the only example in his work and it brings home even more forcibly the concept he used in all his three-hinged arches i.e. that the bridge is slenderest at the points one would expect it to be thickest. Students and engineers will excuse my ignorance of stresses but that is how it strikes me! The reason for the bad quality of the photograph is that, although it is possible to see the bridge quite well from downstream, its upkeep has been badly neglected. Two ugly poles droop from the centre almost to water level and piles of lumber and bric-a-brac are stored under its fine aristocratic spans.

An abused and forgotten masterpiece indeed.

(Map Coordinates: 46.5929309, 7.3753918)

6. Schangnau – The Hombach Bridge (1931)

The first of the two Schangnau Bridges I came to, driving up over the rolling hills from Thun, is a simple, low-lying structure with a light-hand railing which makes it more "springy" than the ones I had just visited.

The little stream it spans has been carefully maintained as can be seen in the photo and the whole setting is neat and clean. The cantonal authorities have done justice to Maillart's work by looking after the environment since the road which it supports carries much more traffic than it did in his day. The bridge is smart and free from debris and the undergrowth has been kept at a healthy distance.

This was the first time I had a real impression that someone really cared about his work.

(Map Coordinates: 46.8178178, 7.8282838)

7. Schangnau – The Lüterstalden Bridge (1931)

The same "cared for" feeling goes with the second Schangnau Bridge about a kilometre down the same road. With its airy railings and simple form, it blends with its wooded surroundings. The trees and saplings around its arch have been cleaned away and I must say it is looking very healthy. The rather massive end walls which saddened Max Bill have now been discretely covered by the earth and rocks cleared from the riverbed.

It seems a pity that such care has been taken to preserve these fine but rather simple bridges when other masterpieces are still lying forgotten – for want of a major road.

(Map Coordinates: 46.8202945, 7.8450724)

8. The Ladholz Bridge
(1930/31)

Off the road from Frütigen to Adelboden, lies one of Maillart's most enchanting bridges built in the same period. This one is a real puzzle for the bridge hunter and took me a long time of map-poring before I located it.

Truly one of Maillart's forgotten bridges, it lies several hundred feet from the highway down a steep and rugged path used only by the occasional fisherman or nature-lover. In fact, as I laboured down the slope a chamois deer darted away through the thick pine forest.

I came upon the bridge unexpectedly when the sloppy mud path suddenly became concrete and I realised I was on the bridge with its solid narrow parapets taking me across the wide bed of the river.

I scrambled down to the river and, as I crouched on the rocks beside the water, the solid imaginative structure seemed to be out of place in the wild and lonely setting. Not because it clashes physically with nature but, on the contrary, because a genius had spent so much time and energy on creating one of his works of art where almost nobody would ever see it.

(Map Coordinates: 46.5356428, 7.5980041)

9. Adelboden – The Spitalbrücke (1930/31)

Just up the road from where I walked down to the Ladholz bridge is the Spitalbrücke which crosses the river Engstiglen a few kilometres before Adelboden. It is quite difficult to find since there are several crossings of the river after Frutigen.

However, there is no mistaking Maillart's style once you clamber down by the ice-cold rushing river. The bridge spans the torrent obliquely giving an interesting twist to the solid slabs of the twin-arched ribs. Maillart was very keen on this solution since, as he says in an article in the SBZ (April 1936), the same shuttering can be used several times which greatly reduces the cost of the scaffolding.

Whatever his reasons for adopting this solution, the result is eye-catching and harmonious in its mountain setting.

(Map Coordinates: 46.5029579, 7.5806321)

10. Habkern – The Traubach Bridge (1932)

I had come to Habkern late in the previous autumn with my friend Tom during my first attempts to find some of the lost bridges. Down in Interlaken, the sun had been shining through the last of the golden leaves but as we had driven up the narrow track towards Habkern the snow started falling and the going got worse and worse. We made the village but the blizzard reduced visibility to nothing. I had only my little "concrete" map to guide me so we decided to call it a day and slid back down to the lakes.

But this time in early summer I could have been on a picnic. I found my way easily from the village down the track towards the Traubach and there was the bridge. It proved difficult to photograph but even so I think this view gives a good impression of this sturdy practical bridge built to carry the herds of cattle down to the village.

(Map Coordinates: 46.7304135, 7.8663233)

11. Habkern – The Bohlbach Bridge (1932)

This second Habkern Bridge is only a stone's throw up the winding track from the Traubach Bridge and is set amidst the trees and luxuriant undergrowth.

Although it is only a very small structure and has rather heavy side walls to stiffen the arch, it represents the first step towards the magnificent Schwandbach Bridge. It carries the road in a sweeping curve movement from one side of the small Bohlbach Valley to the other so that it can continue its winding way to Schwendi. Few that cross it possibly ever pause to consider the artistry which went into making this bridge. In fact it blends so well with nature that you scarcely notice it. Maillart himself would have chuckled to see his little prodigy so forlorn and overgrown.

Max Bill prophesied that, on the centenary of Maillart's death, all his bridges would be cleaned and painted in honour of their maker. Alas, the little Bohlbach Bridge in 1973 has lost all pretentions to glory and is content to let the branches and bushes tenderly entangle it.

(Map Coordinates: 46.7323283, 7.8698544)

Habkern – The Bohlbach Bridge (1932)

12. Zweilütschinen –
The Gündlischwand Bridge (1937)

A strange and rather haunting bridge this and almost unnoticed from the road. It crosses the river obliquely and thus does not alter the direction of the roadway. It is only when you crouch in the thick flower-studded grass along the river bank that you notice the curious construction. It is a double girder bridge where Maillart used the same aesthetic moulding as in several of his bridges over railways.

However, because of its low-slung construction the pier is no longer visible. He has used the same moulded parapet as with the Huttwil and Liesberg Bridges but built so near the water it gives it a completely different aspect.

A unique feature of the bridge is the small lip which juts from the parapet facing upstream. An artistic quirk to throw a curving shadow? A structural necessity? I wonder if Maillart foresaw the phenomenon so evident in the photo; a clean, stain-free rectangle cutting the bridge into three!

(Map Coordinates: 46.6317519, 7.9121703)

13. Innertkirchen – The Aare Bridge (1934)

I had spent the night in Interlaken and alas the rain was already in the air as I drove quietly along the banks of the second of the two lakes which have given the town its name. By the time I got to Innertkirchen it was raining heavily as can be seen in the photograph. I was disappointed by the rain and also by the bridge which had looked so beautiful in Max Bill's book. The little town has grown up around the approach road and the albeit welcome vegetation now makes the bridge almost impossible to photograph from the "right" side.

I say the "right" side because the upstream side of the bridge has been altered by the addition of a pedestrian walk-way which has completely changed the aspect of this side. It is only when you climb down under the arch that you can see that the basic smooth lines are still there. A downstream view still gives the sweeping slenderness that Maillart wanted to portray but from no viewpoint can you see it as he intended.

(Map Coordinates: 46.7054022, 8.2280098)

14. Wyler – The Gadmerwasser Bridge (1938)

As if to lift my spirits, the sun broke through the rain clouds as I climbed out of Innertkirchen in search of the strange Wyler Bridge which no one but my "cement" map had ever mentioned. When I eventually found the classic masonry arch, completely hidden from all sides by encroaching vegetation, I realised that even Maillart could not have been very proud of this one!

I almost slipped and hurt myself in the descent of the precipitous sides of the ravine over which the bridge was suspended and once there found I could not even see the structure for the trees and creepers. Eventually, having scrambled in every direction, and soaking myself into the bargain, I caught a brief sunlit glimpse of half an arch and snapped what is the only view possible of the Wyler Bridge.

Maybe I should return in the winter – at least the trees will not block the view.

(Map Coordinates: 46.7087603, 8.2427857)

Wyler – The Gadmerwasser Bridge (1938)

15. Nessental – The Triftwasser Bridge (1932)

Of all Maillart's bridges the simple beam bridge over the Triftwasser Gorge is the most delightfully situated for the rambler and nature-lover. Leaving the main Sustenpass Road just after Nessental, you take a small track that leads off to the right. A few hundred metres on this steep and twisting farm road and you come upon a small chalet and a woodman's sign post. It is only a short walk from the chalet through lush meadows and fragrant pine forest to the bridge.

Heady and happy with the odour of the pines and the sweetness of the air, it was over an hour before I could pull myself away from the forest and continue my search. In all that time, not a single soul set foot on the bridge and possibly would not for the rest of the day. Only the birds and the squirrels could share the hidden treasure I had found.

(Map Coordinates: 46.7170055, 8.3217131)

Nessental – The Triftwasser Bridge (1932)

16. Ibach – The Muota Bridge
(1912–13)

My next stop being Schwyz and the Muota Bridge, I had planned to cross the Susten Pass and drop down through Wassen and Altdorf. However, the pass was still closed to traffic despite the fact that June was just around the corner. So I had to retrace my steps and take the long way round via Luzerne.

The Muota Bridge in the little town of Ibach was alive with people and traffic. A good deal of water had to flow under it before I managed to take the photo I wanted – with no traffic and just the simple cross set on its parapet. A second after my shutter clicked a great articulated lorry lumbered across. At least I had found a bridge that really served its purpose, despite its rather classic and heavy unimaginative construction. Technically, it is a very low bridge with cantilevers at each end and a beam suspended between. The effect from the unencumbered banks is not unpleasant especially since the addition of a light, railed pedestrian walk has (unusually) given a simpler aspect than was originally intended and also covered the unnecessarily heavy "castle-like" sidewall decorations.

(Map Coordinates: 47.0110670, 8.6443720)

17. Altendorf – The Railway Bridge (1939)

The rain of the previous day in Innertkirchen was completely forgotten as I glimpsed the Lake of Zurich and drove down towards Altendorf and Lachen. The sun beat down and so I opened the top of my convertible Fiat.

In Altendorf I found the first of the railway bridges with Maillart's characteristic "moulded" style. It is a continuous beam construction and represents the pureness and aesthetic simplicity of his work in concrete. These moulded beam bridges always give me the impression of elasticity as if the whole thing had been cast in one rubber piece.

Although the bridge serves only a very minor road going from the village to the lake, the railway setting has ensured that its pureness is guarded against encroaching time. Thus it still stands out clear and clean proudly bearing its "made by Maillart" stamp.

(Map Coordinates: 47.1936697, 8.8309568)

18. Lachen – The Railway Bridge (1940)

In Lachen is the last of Maillart's bridges – built in 1940, the year of his death. So…Maillart went out with a bang and not a whimper!

How disappointed I am with my photos but what a magnificent bridge. Its intricately designed hollow arches literally fly across the rails. Because of the oblique crossing of the tracks these slender arches spring from their abutments at different angles so that, from each new vantage point, the bridge's taut lines change and interweave.

The small photo here was taken from the tracks themselves after I had asked an old lady if I could go through her garden to take a picture. She must have thought me crazy but smiled and opened the latch.

I'm pretty sure that neither she nor the hurried travellers who speed past on steel or rubber wheels realise that beneath their gaze lies the ultimate symbol of Robert Maillart's life work. I am pleased that despite frustrations and disappointments he was able to end his career and his life with such a fitting monument.

(Map Coordinates: 47.1901416, 8.8395988)

19. Innerthal – The Schräbach Bridge (1924)

Originally there were two bridges like the Schräbach in Innerthal but one was destroyed and replaced by a new and "modern" structure.

I find this a bit ironic because, despite its present appearance, the system used for the Schräbach and Flienglibach Bridges was the avant-garde stiffened-arch system to which Maillart dedicated his life's experiment. Originally this strange, grey bridge was a light and willowy construction based on Maillart's stubbornly-held theories of how weight reduction combined with aesthetic beauty retains strength.

Max Bill recounts that shortly after they were built the two bridges suffered frost damage and the reinforcing steel threads started breaking out of the poorly mixed concrete (no fault of Maillart's design). Repairs were made by spraying liquid concrete onto the defects but for some reason someone decided that the sides must be walled in too. Thus a beautiful bridge became a heavy eyesore which (I don't know why) reminds me of something out of a prisoner-of-war camp.

(Map Coordinates: 47.1044516, 8.9012224)

20. Wägitalersee –
The Ziggenbach Bridge (1924)

A few miles round the wooded shores of Lake Wägital lies the delightful little curved bridge over the Ziggenbach. What a contrast to the sorrowful Schräbach Bridge which stands in solemn sentinel before the pines.

This forgotten masterpiece is one of my favourites. It has happily been left as it was built and curves the road round over its simple fixed arch in 5 segments. The setting is quite breath-taking in that, on one side, the steep river ravine climbs up towards sculptured mountain peaks and, on the other, drops down in little sparkling waterfalls to the deep blue lake.

Again, this bridge is difficult to photograph since time and neglect have clustered trees and undergrowth round the arch and approaches. The quiet serenity of the Ziggenbach Bridge and its surroundings held me in its spell for many moments and when at last I drove away, I secretly hoped that it would remain forgotten and hidden so that I could return and rediscover it.

(Map Coordinates: 47.0775349, 8.9290671)

Wägitalersee –
The Ziggenbach Bridge (1924)

21. Wattwil – The Thur Bridge
(1909)

I enjoyed my visit to this old and classical bridge in Wattwil despite its heavy decorations and what I call "pseudo city" style.

It was Sunday and the faithful folk were flocking to the church across the water pausing on the bridge to gaze down pensively as people will. The deep church bell echoed, calling them to prayer across the quiet town and I felt like an intruder crouched amid the flowers on the river bank in my old jeans and mud-spattered boots.

An old fisherman greeted me gruffly as he hobbled by with line and tackle relieving me of my guilt; so I lay back in the grass and watched the old bridge and the bobbing heads of the pilgrims.

People come and people go, treading their way across the river. Their days are numbered and when at last they make that final one-way trip to the church across the water, the bridge will remain, strong and silent for their children and their children's children and life goes on…

(Map Coordinates: 47.3019931, 9.0862321)

Wattwil – The Thur Bridge (1909)

22. Schiers – The Salginatobel Bridge (1929/30)

When I finally found the Salginatobel Bridge, it was one of the most exciting discoveries of my life. This is undoubtedly the most imposing, and the most beautiful bridge Maillart ever built. The feelings I experienced when I caught my first glimpse of the forgotten masterpiece made me realise exactly why I had undertaken my Bridge Hunt in the first place. As I explained in the first chapter of this book, I had already tried in vain to find the Salginatobel Bridge the year before and had come tired and dispirited to within half a mile before turning back.

And now here I was at last, alone in the rugged valley feasting my eyes. I spent the better part of the day scrambling around trying to get the photograph I wanted. I drove over the bridge (which is just wide enough for one car) and continued for several kilometres up into the mountains trying to find a better vantage point but to no avail.

In the end, I decided that the picture I wanted could not be taken from the Salgina Gorge nor from the immediate vicinity of the bridge and drove back down the narrow winding farm road to Schiers. I had spent several hours contemplating the beauty of the setting and in all that time only one old Volkswagen from the hamlet up in the mountains had crossed the bridge over the Salgina Valley. As I drove down, I noticed a tractor pulling a load of timber way over on the other side of the Schraubachtobel Valley (the Salgina Valley runs into this one) and immediately realised that from where the tractor was, I could probably get an even better view of the bridge.

I had never thought of taking a completely different road in order to photograph the bridge so I stopped to study the map.

*Schiers – The Salginatobel Bridge
(1929/30)*

Sure enough, there was a small dotted line on the map going along the edge of the valley opposite to where I stood. But it was hardly a road – probably a mud track for the lumber tractors to bring their great tree trunks down to the village. However, back in Schiers I decided to risk it and, in my decidedly un-tractor-like car, I nosed my way up the even more formidable mountain road on the other side of the valley.

After about three kilometres I saw a mud-track leading off to the left and, consulting the map, I saw that this was the dotted line where the tractor had been. Unfortunately, an imposing sign marked the entrance to the little track threatening all sorts of undesirable consequences to those unauthorised persons who ventured upon its muddy way.

I figured that sometimes bridge hunters are obliged to take risks and, having almost injured myself several times already in perilous descents, I decided that a good picture was worth most of the penalties the Swiss authorities could pile on me. So I gingerly inched my way into the lane.

About a kilometre up the track, I heard a thunderous noise approaching and squeezed my small car into a clearing under the pine trees just in time, for an enormous tractor (probably the one I had seen from the other side of the valley) came lumbering past, with inches to spare, between my bright paintwork and its tree-laden trailer. So then I decided to walk!

It was getting late and it was only twenty minutes further on, with the light going fast, that I saw what I had come for.

There, between the pines, hung like a distant thread was the Salginatobel Bridge. From my vantage point, I was looking across the Schraubachtobel Valley and up into the Salgina Gorge so that my gaze caught the bridge side on. Immediately, I realised that this was the photo that would adorn the cover of my book – if I ever got that far.

I heartlessly snapped a few photos in the descending twilight but told myself I had to return the next day.

The following morning, I left my hotel in Vaduz (Liechtenstein) at about 05h00 and drove back to Schiers which I reached just as dawn was breaking. Without a second thought, I sped through the sleeping village, up the winding mountain road and onto the forest track. I drove straight to my view point and clambered up a steep embankment.

It was 0700 and wisps of clouds were clinging to the wooded slopes of the mountains making the whole scene rather mysterious and awesome. I am pleased to say that I got the picture I wanted and after a lingering moment drinking in the early morning calm and beauty of the exquisite Salginatobel Bridge, I made my way down to the valley to continue my quest for forgotten bridges.

(Map Coordinates: 46.9817702, 9.7181357)

23. Klosters – The Landquart Bridge (1930)

The railway bridge over the river Landquart in Klosters presents a completely different picture of Maillart's pioneering work in reinforced concrete from the Salginatobel Bridge I had left that morning.

This very elegant, arched-rib structure, continues the curve of the railway, as it comes out of the tunnel on the west side of the river into the station on the east side.

The arch itself tapers to half its width at the summit and, viewed from the side, you can almost feel the cushioning of the various stresses absorbed by its graceful form. The suspended footbridge, an ingenious device, was added later to prevent damage by earth movement which threatened the bridge. In fact, the weight of the footbridge pulling on the suspending wires serves as a strengthening factor to take up the extra new stresses on the concrete supports.

(Map Coordinates: 46.8676995, 9.8807517)

Unfortunately, both the Klosters Town Hall and the local railway company inform me that this bridge was demolished in 1993 to make way for a warren type truss bridge – to my eyes a heavy and ugly replacement.

24. Zuoz – The Inn Bridge (1901)

I have seen the Zuoz Bridge quoted as being Maillart's first bridge. In fact, it was his second – built two years after his Stauffacher Bridge in Zurich.

However, it certainly was the first bridge of the "Maillart system" and was constructed as a three-hinge, box-section arch. This was my second visit to Zuoz, since I had come here in the previous autumn and discovered my first Maillart bridge as I described at the beginning of the book. I was again moved by this simple, elegant structure with its clean lines unchanged for 72 years. The old photos in Max Bill's book taken at the time of construction could have been snapped today – nothing has changed.

Just as Maillart finished his bridge-building career with a masterly performance (the Lachen Bridge) I think you could say that the Zuoz Bridge represented a pretty good debut!

(Map Coordinates: 465969303, 9.9627613)

25. Donath – The Val Tschiel Bridge (1925)

I did not expect to be as enchanted with this lovely bridge as I was. It spans a picturesque ravine hidden away in a remote corner of the Canton of Graubünden.

The quaint Swiss village of Donath should be proud of their collector's item for it has preserved all the dignity and simplicity of its original form.

As in the Schangnau Bridges, the rather heavy-end walls which must have spoilt the aesthetic effect of the bridge when it was built, are now discretely hidden by vegetation and scree. The result after nearly 50 years is a joy to behold. Although the bridge is well hidden off the beaten track, it is well-worth a visit even by the layman. If nothing else, the grassy slopes dropping suddenly into the river gorge winding below make an excellent site for a picnic.

(Map Coordinates: 46.6285038, 9.4259607)

26. St. Gallen – The Steinach Bridge (1903)

Surprisingly enough, this strange bridge set in the town centre with buildings on every side and even underneath it, was very difficult to find. I had misunderstood the brief instructions on my "cement" map and was circling the sprawling town for a long time before I found the bridge.

Of course the classic style and ornate railings of the bridge were not the Maillart style I was used to, but the site could not but intrigue me. The little stream over which the bridge spans is almost lost from view now among the cluster of buildings which flank the small valley.

While talking to a Swiss friend recently about my project, I mentioned this curious St. Gallen Bridge.

When I outlined where it was, he exclaimed, "Ah, yes, I remember, we used to go drinking as students in a little bar right under that funny bridge!"

(Map Coordinates: 47.4216687, 9.37800658)

*St. Gallen – The Steinach Bridge
(1903)*

27/28. Aach – The 2 Railway Bridges (1907)

Here in the tiny hamlet of Aach, set among the wide fields which stretch down to Lake Constance near Romanshorn, I found the two earliest examples of Maillart's "moulded" style of beam bridges.

The photograph shows the two identical bridges which carry one-lane farm roads over the railway lines. The moulded style has by no means reached the stage of elasticity which Maillart achieved in Liesberg or in Altendorf, but nevertheless possess a remarkable simplicity and beauty for the period in which they were built.

(Map Coordinates: 47.5512950, 9.3465037)

29. Oberüren – The Thur Bridge (1903/4)

This very old Maillart bridge takes the road from Oberbüren (just off the Zurich-St. Gallen auto route) to the little village of Bilwil. To me, the only particular feature of this bridge is the wonderful sense of calm and timelessness you feel on the quiet river banks. By dint of patience and chance, I even managed to snap a fish jumping from the still waters on that sunny afternoon as I sat gazing at the bridge.

So hypnotic was the silence and tranquillity bestowed by the scene, that I didn't even notice the man on the bridge in my photo until long after when I was projecting my slides. And even then I had to go up to the screen and touch it before I would accept that he had really been there.

(Map Coordinates: 47.4597051, 9.1638027)

30. Falsegg – The Thur Bridge (1933)

As can be seen from this photo, I had come this way before. Although, on my second visit to this graceful bridge I also took some early summer photos, I felt that this autumn shot was better. The slender pointed arches which Maillart used here for the first time lend a certain "animal" movement to the style we now know so well. Some articles I have read (and there are not many) have likened these 3 hinged-arch bridges to a greyhound and I think the analogy is a good one.

The long approach road supported by thin concrete pillars give a kind of exciting acceleration to the structure to prepare it like a graceful animal for its final leap across the low, wide river. Again, the incredible fineness of the arches, as they reach the ground, defies my layman's imagination when I think of the forces pressing down.

As a note for visitors, this bridge is clearly visible from the St. Gallen-Zurich autoroute so that for once some of Maillart's finest work can be appreciated by all.

(Map Coordinates: 47.4635182, 9.1142413)

The Thur bridge as it is in 2022 – seen from the autoroute.

31. Winterthur – The Toss Footbridge (1934)

This charming, modern-looking footbridge in the suburbs of Winterthur (built in 1934!) ranks among my favourite Maillart bridges.

Here, he was able to allow all his creative artistic forces to come into play – untrammelled by considerations of heavy loads and durable road surfaces. And how magnificently he succeeded.

The bridge was designed in collaboration with W. Pfeiffer, a Winterthur engineer, and in my mind, could not be surpassed for sheer elegance. The line of the curve which seems fairly flat when viewed from downstream, arches up into a supple sculpture when seen from the roadway.

As Max Bill enthuses: "It has the ease and pleasant naturalness as if it had grown there and sought to span the river itself."

Although the town has spread out its paved tentacles all around the bridge, the view from the banks of the river has retained its original charm. Mothers and children and quiet folk exercising their dogs wander over the bridge and I think that Maillart would have been content to see how pleasantly useful it has become.

(Map Coordinates: 47.5023378, 8.6910780)

One feature I appreciate in this structure is the low, side walls supporting the simple railings which sweep around at either entrance to the bridge, lending to its suppleness.

As a postscript to the Winterthur footbridge, I include this snapshot taken on a recent trip through the centre of Iceland. On a crushed lava track over the one of the many salmon-packed rivers, stands this strangely familiar bridge. It is located near the tiny hamlet of Nes in North Iceland (near Akureyri) and I cannot help but think that, at least in Iceland, someone had heard of Robert Maillart!

The Maillart "look-alike" bridge in Iceland.
(Map Coordinates: 65.7256671, 17.9024517)

The footbridge as it is today

32. Zurich–The Stauffacher Bridge (1899)

This, the first ever Maillart bridge, stands solid and ornate in the centre of Switzerland's busiest city. Here it was that the 27-year-old Robert Maillart, fresh from Zurich's world-renowned Technical University, started his life's work and experimentation in reinforced concrete. How sad that the architectural modes of his youth dictated the decorations and adornments which have completely hidden the first bridge of the Maillart system.

As a layman, the Stauffacher looks to me like any other classical heavy city bridge. Its four-posted lions and dated wreath do not jar the eye so used to "pseudo city" architecture and we can only wonder at Maillart's genius when, gazing at the 1901 Zuoz Bridge, we realise what an adventurous and courageous step he took only two years later.

(Map Coordinates: 47.3711289, 8.5308782)

Laufenburg – The Rhine Bridge (1909)

33. Laufenburg – The Rhine Bridge (1909)

The two Rhine Bridges, among the largest of all Maillart's bridges, are notable only for their picture post-card settings.

Constructed without hinges from concrete and granite blocks, I think of them as bread-and-butter bridges. In their functionality we cannot fault them but we know that, had the young Maillart had his way, we may have seen slender leaping structures spanning the Rhine in these two lovely towns.

To be fair however, the bridges do harmonise magnificently with the medieval charm of the ancient townships. Here in Laufenburg the heavy old bridge is somehow what we expect to see and maybe a more modern structure would have hurt the eye. Maillart or no Maillart, Laufenburg is a delightful spot where the 20th century has had great difficulty in permeating its narrow streets steeped in history.

(Map Coordinates: 47.5638305, 8.0609927)

34. Rheinfelden – The Rhine Bridge (1909)

Despite its heavy unimaginative style, this is the longest bridge Maillart built, spanning the ever-widening Rhine by jumping on and off a little tree-covered island.

In order to take this picture, I left my car in the medieval streets of the Swiss town and walked across the bridge, passport in hand, to Germany! The German guard listened patiently when I told him my mission and let me go.

He seemed surprised when I returned 30 minutes later thanking him and telling him everything was OK.

(Map Coordinates: 47.5549722, 7.7902197)

35. Liesberg – The Birs Bridge (1935)

There is no doubt that this is the most difficult of Maillart's bridges to find. If any future bridge hunters wish to gaze upon this perfect example of a continuous beam "moulded" railway bridge, I wish them luck! Even though I had a very large-scale map with me, I almost went away convinced that it had been destroyed. This hidden masterpiece is buried beneath a jungle of vegetation within the fenced confines of a cement factory.

Making one last desperate attempt to find it, I crossed two fields of waist-high grass and climbed gingerly through two electrified fences into a dense tangle of stinging nettles. Smarting and sweating, I reached the muddy river bank. Nothing was visible, so I hacked my way to the barbed wire factory fence jutting right down into the dirty water of the Birs.

I managed to clamber over by climbing a tree and there, appearing through the undergrowth, was what looked like a bridge. I discovered on closer inspection that its piers were in the water – the only Maillart bridge to get its feet wet – and scrambling up into the cement factory yard, I found that it only served a seemingly disused branch railway line.

The parapets have that appealing moulded aspect I found in Zweilutschinen and there is a very pleasing harmony about the whole structure. Although no one hailed me in my hideaway down by the bridge I had a guilty feeling of trespassing as I slipped away back to the road through the undergrowth.

(Map Coordinates: 47.4000928, 7.4408853)

This Photo is of the Birs bridge today – now visible from the road which passes through the village of Liesberg.

Aarburg – The Aare Bridge (1911/12)

36. Aarburg – The Aare Bridge (1911/12)

Only the arch of this fine bridge is left from Maillart's original design. The supporting members which once connected the platform to the arch have been removed and the roadway is now supported by new beams. The bridge is built on a gradient of 5% and its "new look" has by no means detracted from its original elegance. It harmonises very well with the picturesque surroundings.

I lingered several hours under the trees beside the calm waters of the Aare, throwing scraps of bread from my sandwiches to the ducks. I am relatively pleased with my photograph of the bridge underneath the castle since I captured just the angle and the atmosphere I wanted.

(Map Coordinates: 47.3203376, 7.8982586)

37. Huttwil – The Railway Bridge (1935)

I found the Huttwil Bridge very difficult to photograph owing to the surrounding paraphernalia and was not helped by the fact that the sun was facing me. However, Huttwil was rather out of the way and meant a long detour so I decided that whatever came out would have to do. The result is satisfactory and gives a good impression of how, once again, Maillart accepted the challenge of spanning a void obliquely in order to preserve the smooth flow of the roadway. Very similar in construction to the Liesberg and other beam bridges, it conveys very well that elasticity of form which pleases me so much.

(Map Coordinates: 47.1145449, 7.8631853)

38. Bern – The Railway Bridge (1938)

Built at the end of the platform of the Fischermatteli station in the suburbs of Bern, this sad bridge has all but disappeared beneath the trappings of railway electrification.

The once clean, pure lines are now barely discernible with station buildings cramped up to and even under the eastern side.

Even so, the eye can still pick out the cool moulded lines of this typical Maillart construction. I have to take Max Bill's word for it when he says that the bridge belongs to the best of Maillart's works, for I came away wet and disappointed!

(Map Coordinates: 46.9421890, 7.4110094)

39. Bern – The Lorraine Bridge (1928/30)

The Lorrainebrucke in the centre of Switzerland's capital is the largest and most massive of all Maillart's bridges. In fact, we can see from the inscription on the bridge itself (photo below) that Maillart was not alone in the realisation of this rather overpowering and classic project. The architect was Hans Klauser and I have heard through various sources that Maillart was not at all happy with the finished bridge.

Nevertheless, once you climb down the steep, wooded slopes beside the bridge, the noise and bustle of the city is left behind and you discover a peaceful tow path beside the deep, still river.

(Map Coordinates: 46.9526296, 7.4428484)

40. Schwarzenburg – The Schwandbach Bridge (1933)

The two beautiful bridges near Schwarzenburg are so well hidden that you really need a precise map and clear instructions if you want to find them. From Schwarzenburg you take the road towards Riggisberg and Kirchenthurnen. A few kilometres from the town is the tiny hamlet of Schönentannen and here you take the minor road which leads through farmland and forest to the bridges.

I have included the famous Schwandbach Bridge first although it is the second of the two bridges you come to. Considered by many experts to be the most beautiful of Maillart's constructions, the bridge spans a deep, river ravine and is preserved like a jewel in the magnificent unspoilt woodland scenery.

In the 40 years since it was built, the forest has crept in, all around the bridge, making it almost impossible to photograph. However, when you see the majestic curving contours of the bridge you can immediately appreciate the combination of Maillart's technical prowess and artistic genius. As with the Salginatobel Bridge, the spanning problem was a difficult one but Maillart not only conquered it in a technically perfect way but with a maximum of economy and beauty. The only thing I can really say about the Schwandbach Bridge is that it has to be seen to be believed.

(Map Coordinates: 46.8293498, 7.4023745)

41. Schwarzenburg –
The Rossgraben Bridge (1932)

A few meters down the woodland road is the elegant Rossgraben Bridge, which leaps the low, wide river bed like a lithe, white animal amid the tall pines.

Here are no heavy parapets or masonry sidewalls to mar the perfect flowing lines of the bridge and you cannot help but think that this must have been one of Maillart's favourite creations.

Although, similar in appearance to the Geneva and Falsegg Bridges, the "bow" of the bridge seems to arch higher and the slender railings add no unnecessary weight. It seems to prove once and for all Maillart's conviction that the strength of a bridge lies in the creative and informed mind of its creator and not in the abundance of materials used.

Schwarzenburg – The Rossgraben Bridge (1932)

42. Twann – The Waterfall Bridge (1936)

Here on the banks of the Bielersee the vineyards slope down to the lake and the picturesque villages abound with flowered gardens alive with roses and geraniums. And it was here that I found the last of the 42 extant Swiss Bridges of Robert Maillart and completed my bridge hunt. The little three-hinged arch bridge above the hamlet of Twann is set over a beautiful waterfall – a delightful spot for the last visit of my long pilgrimage.

The bridge was built with solid sidewalls because the Cantonal authorities dictated that a heavy solid construction was needed in order to blend the bridge into the surrounding nature. How strange we consider this attitude nowadays and how sad that Maillart was not able to show them how beautifully he could have solved the problem.

But neither sorrow nor regret were in my mind as I gazed upon what was to be the last of my Maillart's bridges, set against the peaceful waters of the beautiful lake.

(Map Coordinates: 47.0933537, 7.1516648)

JOURNEY'S END
The quest for all these hidden masterpieces has led me on a magnificent journey through the mountains and valleys of all Switzerland and I feel a sense of gratitude for the wonderful things Robert Maillart has led me to see.

Twann – The Waterfall Bridge (1936)

Afterword

....and the first shall be last!

And now, 50 years later, I'm still standing and so are the 43 bridges (with the exception of the Klosters bridge and the totally altered Vougy bridge)

Yes .....43!

As mentioned in the Foreword, while fact-checking my 1973 manuscript I happened to be skimming through Professor David Billington's 1979 book "Robert Maillart's Bridges - the art of engineering" (Princeton University Press) and my eye was caught by the following paragraph:

"Pampigny: The First Bridge

As Maillart's studies drew to an end in March of 1894, he found work with the firm of Pumpin and Herzog in Bern. In October of 1895 the firm sent Maillart to Morges, a town just west of Lausanne on Lake Geneva, where they had built a private rail line for a company known as B.A.M. (Bière-Apples-Morges). The section from Morges through Apples to the village of Bière had been opened in June and Pumpin and Herzog were to build a branch line from Apples through Pampigny to the village of L'Isle. Since Maillart had evidently worked in the Bern office on the layout of the line, as well as on the design of its bridges, the firm sent him to the branch site as an assistant to the supervising engineer. Maillart stayed in Morges until March 1896, when he moved to Pampigny. Here, between March and September he supervised the construction of his first bridge, having worked on the plans in late 1894"

So …. the Stauffacher bridge was really his second one and the Zuoz bridge his third!

My guess is that Maillart, as a graduate trainee with Pumpin and Herzog, was probably not entirely responsible for the design of the Pampigny bridge… but even so!

Further research has enabled me to track down a photo taken in 2011 by the late architect Christian Menn.

This is the Pampigny bridge over the Veyron brook. The railway line is no longer private but still operates … as indeed does the bridge. Not a masterpiece but a collector's gem nevertheless. The map coordinates for bridge hunters is:

(Map Coordinates: 46.594646, 6.405674)

In conclusion, I've managed to glimpse most of Maillart's readily accessible bridges as they are today on Google Maps and I hope that the less accessible ones have survived due to sheer lack of traffic. I guess it's up to other bridge hunters to prove me wrong!

I have thoroughly enjoyed re-looking this manuscript and was able to live again my youthful journey while digitally rebuilding its memory-laden pages.

As a respectful post-script to my attempt to share my admiration for Maillart's creative genius, here are my favourite five bridges as seen by the Google van in 2022.

Wagitalersee
The Ziggenbach Bridge

Salginatobel

The Winterthur footbridge

Falsegg Thur Bridge

Donath – Cleaned then by-passed... but not surpassed

Google Maps show me that 50 years later this lovely bridge
has been cleaned but by-passed and sits unused like a silent
wallflower beside its successor.

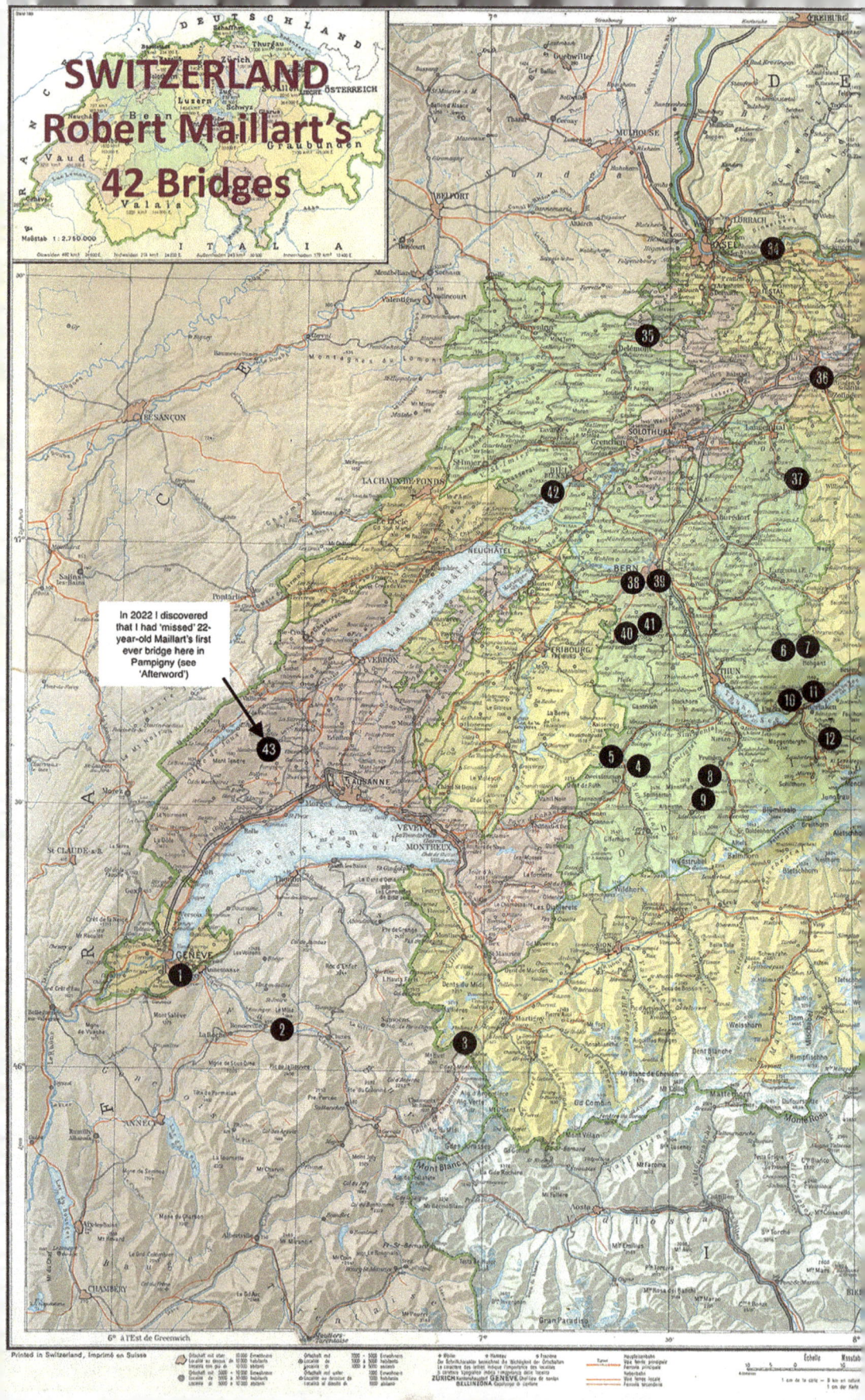

SWITZERLAND
Robert Maillart's
42 Bridges

In 2022 I discovered that I had 'missed' 22-year-old Maillart's first ever bridge here in Pampigny (see 'Afterword')

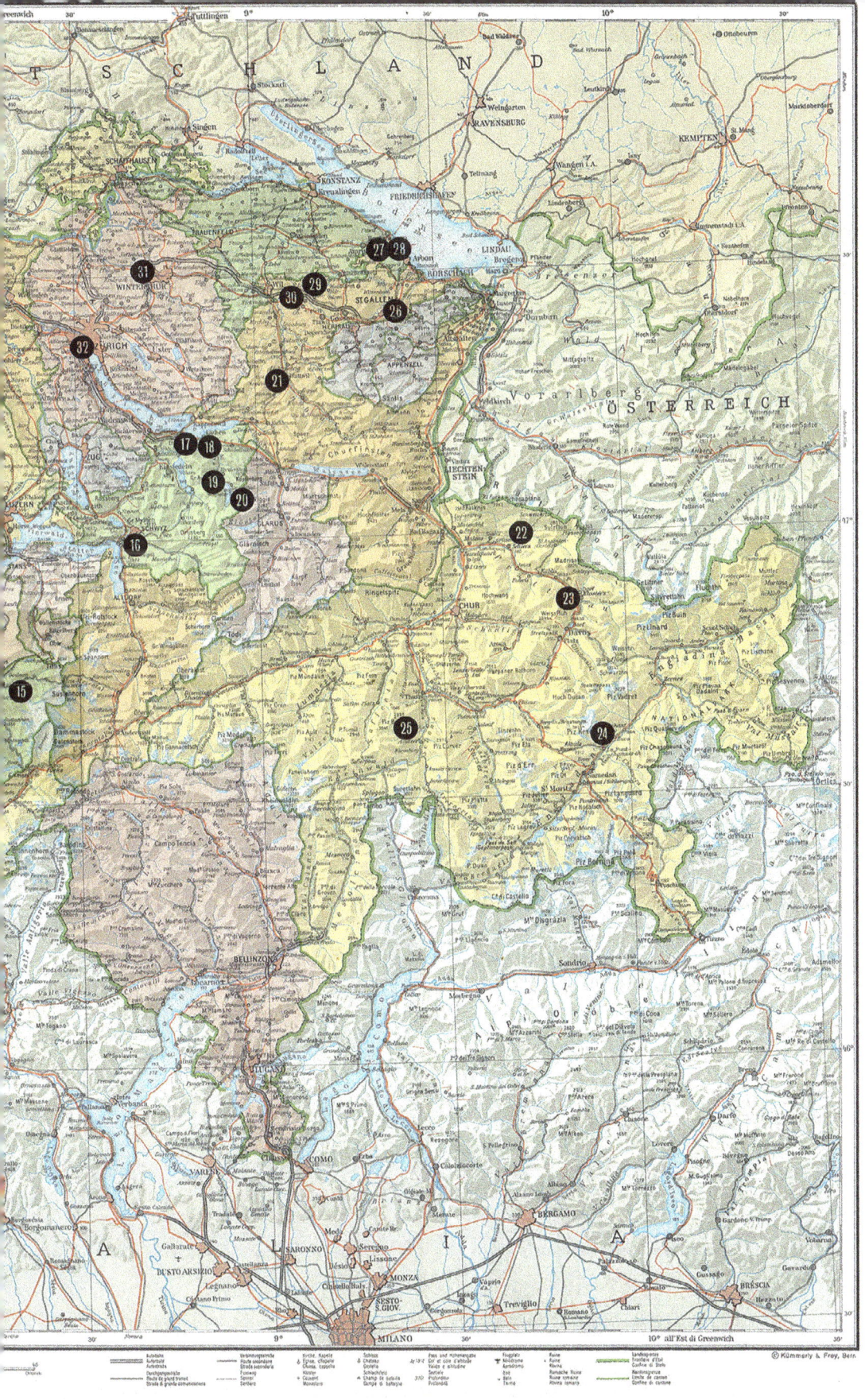